Algebra: A Complete Course

Second Edition

Module A – Progress Tests

Written by Larry Collins

VideoText *Interactive*

Algebra: A Complete Course
Module A – Progress Tests, Second Edition

Send all inquiries to:
VideoText*Interactive*
P.O. Box 19761
Indianapolis, IN 46219

ISBN 1-887891-67-6
6 7 8 9 10 - RPInc - 04

Table of Contents

Progress Tests

Unit I – The Structure of Mathematics

Quiz Form A

Name ______________________

Class ________ Date ________ Score ______

Unit I – The Structure of Mathematics
Part A – Mathematics as a Language
Lesson 1 – Mathematical Parts of Speech

Tell what part of mathematical speech each of the following is.

1. ( ) ____________________

2. m ____________________

3. $\neq$ ____________________

4. $\div$ ____________________

5. □ ____________________

6. 88 ____________________

7. $\cdot$ ____________________

8. [ ] ____________________

9. $>$ ____________________

10. 18.7 ____________________

Quiz Form B

Name ______________________

Class __________ Date __________ Score ______

Unit I – The Structure of Mathematics
Part A – Mathematics as a Language
Lesson 1 – Mathematical Parts of Speech

Tell what part of mathematical speech each of the following is.

1. 101.01 ______________________

2. $<$ ______________________

3. $\{\ \}$ ______________________

4. $\div$ ______________________

5. 13 ______________________

6. $\triangle$ ______________________

7. x ______________________

8. $\ngtr$ ______________________

9. $-$ ______________________

10. $(\ )$ ______________________

Quiz Form A

Name ______________________

Class __________ Date __________ Score ______

Unit I – The Structure of Mathematics
Part A – Mathematics as a Language
Lesson 2 – Mathematical Expressions

Tell whether each of the following expressions is an open phrase, closed phrase, open sentence, or closed sentence.

1. $5-3=6\cdot 2$ ______________

2. $9x+4$ ______________

3. $\sqrt{y+4}$ ______________

4. $x-6=3$ ______________

5. $6y+2=50$ ______________

6. $9-6=3$ ______________

7. $3\cdot 7+6$ ______________

8. $(9-5)8$ ______________

Name ______________________________

Unit I, Part A, Lesson 2, Quiz Form A
– Continued –

In the following exercises, take the appropriate mathematical action with each expression. If you need a domain, use $\{4, 5, 6\}$. If you need a replacement set, use $\{10, 11, 12\}$.

9. $2x + x$ ______________________________

10. $5x + 9 = 59$ ______________________________

11. $\frac{13+9}{2} > 5 \cdot 3$ ______________________________

12. $\frac{x}{2} < 13$ ______________________________

Quiz Form B

Name ______________________

Class __________ Date __________ Score ________

Unit I – The Structure of Mathematics
Part A – Mathematics as a Language
Lesson 2 – Mathematical Expressions

Tell whether each of the following expressions is an open phrase, closed phrase, open sentence, or closed sentence.

1. $(7+2)\cdot 3$ ______________

2. $6\cdot 2=7+5$ ______________

3. $y+4=11$ ______________

4. $6x+2=14$ ______________

5. $7-2\cdot 3$ ______________

6. $2y-8$ ______________

7. $\sqrt{y-3}$ ______________

8. $2+4=7$ ______________

Name ______________________________

Unit I, Part A, Lesson 2, Quiz Form B
- Continued -

In the following exercises, take the appropriate mathematical action with each expression. If you need a domain, use {4, 5, 6}. If you need a replacement set, use {10, 11, 12}.

9. $\frac{x}{2} > 3$ ______________________________

10. $3x - 2$ ______________________________

11. $4x - 3 = 41$ ______________________________

12. $6 \div 2 < \frac{10 - 5}{5}$ ______________________________

Quiz Form A

Name ______________________

Class ________ Date __________ Score ______

Unit I – The Structure of Mathematics
Part A – Mathematics as a Language

Lesson 3 – Translation of Mathematical Symbols

Translate the following English phrases into expressions with mathematical symbols.

1. x divided by 7, decreased by 2

2. the sum of x and y

3. 15 less than the total of x and y

4. the difference between a and b, multiplied by the product of x and y

Translate the following English sentences into expressions with mathematical symbols.

5. The product of 4 and a number is 20.

6. 14 plus, 3 times a number is 59.

7. 25 is less than the quotient of 9 and x.

8. The sum of 20 and the product of 9 and x is greater than 110.

Quiz Form B

Name ______________________

Class ________ Date ________ Score ______

Unit I – The Structure of Mathematics
Part A – Mathematics as a Language
Lesson 3 – Translation of Mathematical Symbols

Translate the following English phrases into expressions with mathematical symbols.

1. 12 more than the difference of x and y

2. the quotient of 7 and 3

3. 32 greater than the sum of a and b

4. the sum of m and n, multiplied by the quotient of 1 and 2

Translate the following English sentences into expressions with mathematical symbols.

5. 17 is the difference between 63 and a number.

6. 34 is greater than the sum of 7 and y.

7. 81 minus, 4 times a number is 21.

8. The difference of 10 and a number is less than 52.

Quiz Form A

Name ______________________________

Class ________ **Date** __________ **Score** ______

Unit I – The Structure of Mathematics
Part B – Further Investigation of Number Symbols
Lesson 1 – The Development of Our Number System

Match the term in column one with the correct phrase or symbol(s) from column two by placing the correct letter in the blank.

______ **1.**	Integers	**A.**	$\{\ \}$
______ **2.**	Uniqueness	**B.**	$\{0, 1, 2, 3, ...\}$
______ **3.**	Real Numbers	**C.**	must get an answer
______ **4.**	Rational Numbers	**D.**	$\{... {}^{-}3, {}^{-}2, {}^{-}1, 0, 1, 2, 3, ...\}$
______ **5.**	Closure	**E.**	$\{x \mid x \text{ is even}\}$
______ **6.**	Irrational Numbers	**F.**	$\{1, 2, 3, 4, ...\}$
______ **7.**	Natural Numbers	**G.**	get only one answer for each combination of numbers and operations
______ **8.**	Whole Numbers	**H.**	set of numbers consisting of both rational and irrational numbers
______ **9.**	Existence	**I.**	numbers that cannot be represented as a ratio of integers
______ **10.**	Set Builder Notation	**J.**	answers must be in the set we are working in
		K.	numbers that can be represented as a ratio $\frac{a}{b}$ where a and b are integers and $b \neq 0$.

Name ______________________________

Unit I, Part B, Lesson 1, Quiz Form A
– Continued –

11. In each of the following sets, tell if **a.** multiplication, **b.** subtraction, **c.** division, and **d.** addition are well-defined operations. If not, give an example to show why not.

$$\{0, 1, 2, 3, 4, \ldots\}$$

a. ______________________________

b. ______________________________

c. ______________________________

d. ______________________________

$$\{1, 3, 5, 7, \ldots\}$$

a. ______________________________

b. ______________________________

c. ______________________________

d. ______________________________

Quiz Form B

Name ______________________

Class ________ Date ________ Score ______

Unit I – The Structure of Mathematics
Part B – Further Investigation of Number Symbols
Lesson 1 – The Development of Our Number System

Match the term in column one with the correct phrase or symbol(s) from column two by placing the correct letter in the blank.

________	**1.** Closure	**A.** $\{\ \}$
________	**2.** Natural Numbers	**B.** $\{0, 1, 2, 3, ...\}$
________	**3.** Existence	**C.** must get an answer
________	**4.** Set Builder Notation	**D.** $\{... {}^-3, {}^-2, {}^-1, 0, 1, 2, 3, ...\}$
________	**5.** Rational Numbers	**E.** $\{x \mid x \text{ is even}\}$
________	**6.** Real Numbers	**F.** $\{1, 2, 3, 4, ...\}$
________	**7.** Uniqueness	**G.** get only one answer for each combination of numbers and generations
________	**8.** Integers	**H.** set of numbers consisting of both rational and irrational numbers
________	**9.** Whole Numbers	**I.** numbers that cannot be represented as a ratio of integers
________	**10.** Irrational Numbers	**J.** answers must be in the set we are working in
		K. numbers that can be represented as a ratio $\frac{a}{b}$ where a and b are integers and $b \neq 0$.

Name ______________________________

Unit I, Part B, Lesson 1, Quiz Form B
– Continued –

11. In each of the following sets, tell if **a.** addition, **b.** division, **c.** multiplication, and **d.** subtraction are well-defined operations. If not, give an example to show why not.

$\{0, 2, 4, 6, 8, ...\}$

a. ______________________________

b. ______________________________

c. ______________________________

d. ______________________________

$\{1, 2, 3, 4, ...\}$

a. ______________________________

b. ______________________________

c. ______________________________

d. ______________________________

Quiz Form A

Name ______________________

Class ________ Date ________ Score ______

Unit I – The Structure of Mathematics
Part B – Further Investigation of Number Symbols
Lesson 2 – Fraction Forms and Decimal Forms
Lesson 3 – Changing Fraction Forms to Decimal Forms

Complete the equivalent form indicated for each of the following fractions.

1. $\frac{7}{9} = \frac{\quad}{45}$

2. $\frac{11}{5} = \frac{\quad}{35}$

Simplify each of the following fractions by rewriting them in factored form and finding all the "ones" you can.

3. $\frac{42}{60}$

4. $\frac{16}{80}$

5. $\frac{51}{17}$

6. $\frac{7}{33}$

Name ___________________________

Unit I, Part B, Lessons 2 and 3, Quiz Form A
– Continued –

Find a terminating decimal, equivalent to each of the following fractions. Show your work.

7. $\frac{5}{4}$

8. $\frac{3}{8}$

Find a repeating decimal, equivalent to each of the following fractions. Show your work.

9. $\frac{5}{11}$

10. $\frac{5}{7}$

Quiz Form B

Name ______________________

Class ________ Date ________ Score ______

Unit I – The Structure of Mathematics
Part B – Further Investigation of Number Symbols

Lesson 2 – Fraction Forms and Decimal Forms
Lesson 3 – Changing Fraction Forms to Decimal Forms

Complete the equivalent form indicated for each of the following fractions.

1. $\frac{5}{13} = \frac{}{78}$

2. $\frac{5}{8} = \frac{}{32}$

Simplify each of the following fractions by rewriting them in factored form and finding all the "ones" you can.

3. $\frac{18}{72}$

4. $\frac{13}{55}$

5. $\frac{55}{121}$

6. $\frac{15}{90}$

Name ________________________________

Unit I, Part B, Lessons 2 and 3, Quiz Form B
– Continued –

Find a terminating decimal, equivalent to each of the following fractions. Show your work.

7. $\frac{6}{5}$

8. $\frac{5}{16}$

Find a repeating decimal, equivalent to each of the following fractions. Show your work.

9. $\frac{7}{9}$

10. $\frac{8}{13}$

Quiz Form A

Name ______________________

Class ________ Date ________ Score ______

Unit I – The Structure of Mathematics
Part B – Further Investigation of Number Symbols
Lesson 4 – Changing Decimal Forms to Fraction Forms
Lesson 5 – Percent

Find a fraction, equivalent to each of the following terminating decimal numbers.

1. .12 **2.** 7.05 **3.** .94101

Find a fraction, equivalent to the following repeating decimal numbers. Show your work.

4. $.\overline{23}$ **5.** $.\overline{3}$ **6.** $.\overline{234}$

For each of the following fractions, find **a.** an equivalent decimal form and **b.** an equivalent percent form.

7. $\frac{13}{100}$ **a.** ________ **b.** ________ **8.** $\frac{13}{25}$ **a.** ________ **b.** ________

Name ______________________________

Unit I, Part B, Lessons 4 and 5, Quiz Form A
– Continued –

For each of the following decimal numbers, find **a.** an equivalent fraction form and **b.** an equivalent percent form.

9. .46 **a.** ________ **b.** ________ **10.** .0021 **a.** ________ **b.** ________

For each of the following percents, find **a.** an equivalent fraction form and **b.** an equivalent decimal form.

11. 43% **a.** ________ **b.** ________ **12.** 4.2% **a.** ________ **b.** ________

Quiz Form B

Name ____________________

Class ________ Date ________ Score ______

Unit I – The Structure of Mathematics
Part B – Further Investigation of Number Symbols
Lesson 4 – Changing Decimal Forms to Fraction Forms
Lesson 5 – Percent

Find a fraction, equivalent to each of the following terminating decimal numbers.

1. .84 **2.** 11.12 **3.** .16707

Find a fraction, equivalent to the following repeating decimal numbers. Show your work.

4. $.\overline{6}$ **5.** $.\overline{47}$ **6.** $.\overline{781}$

For each of the following fractions, find **a.** an equivalent decimal form and **b.** an equivalent percent form.

7. $\frac{87}{100}$ **a.** ________ **b.** ________ **8.** $\frac{17}{50}$ **a.** ________ **b.** ________

Name ______________________________

Unit I, Part B, Lessons 4 and 5, Quiz Form B
– Continued –

For each of the following decimal numbers, find **a.** an equivalent fraction form and **b.** an equivalent percent form.

9. .84 **a.** ________ **b.** ________ **10.** .073 **a.** ________ **b.** ________

For each of the following percents, find **a.** an equivalent fraction form and **b.** an equivalent decimal form.

11. 67% **a.** ________ **b.** ________ **12.** 16.8% **a.** ________ **b.** ________

Quiz Form A

Name ____________________

Class ________ Date ________ Score ______

Unit I – The Structure of Mathematics
Part B – Further Investigation of Number Symbols
Lesson 6 – Primes, Composites, and Factoring
Lesson 7 – Least Common Multiple
Lesson 8 – Greatest Common Factor

Find all factored forms and the prime factored form (PFF).

1. 40 ______ ______ ______ ______ ______ PFF ______

2. 39 ______ ______ ______ ______ ______ PFF ______

3. 36 ______ ______ ______ ______ ______ PFF ______

Find the Least Common Multiple (LCM) for each of the following, using the method indicated.

4. By listing the multiples of each LCM ______________

7 ____________________

8 ____________________

5. By factoring each number as a product of primes LCM ______________

18 ____________________

45 ____________________

Name ________________________________

Unit I, Part B, Lessons 6, 7, and 8, Quiz Form A
– Continued –

Find the Greatest Common Factor (GCF) for each of the following, using the method indicated.

6. By listing the factors of each GCF ______________

20 ____________________________

15 ____________________________

7. By factoring each number as a product of primes GCF ______________

36 ____________________________

42 ____________________________

Quiz Form B

Name ______________________

Class ________ Date ________ Score ______

Unit I – The Structure of Mathematics
Part B – Further Investigation of Number Symbols

Lesson 6 – Primes, Composites, and Factoring
Lesson 7 – Least Common Multiple
Lesson 8 – Greatest Common Factor

Find all factored forms and the prime factored form (PFF).

1. 32 _____ _____ _____ _____ _____ PFF _____

2. 27 _____ _____ _____ _____ _____ PFF _____

3. 81 _____ _____ _____ _____ _____ PFF _____

Find the Least Common Multiple (LCM) for each of the following, using the method indicated.

4. By listing the multiples of each LCM ____________

11 ____________________________

3 ____________________________

5. By factoring each number as a product of primes LCM ____________

20 ____________________________

25 ____________________________

Name ___________________________

Unit I, Part B, Lessons 6,7, and 8, Quiz Form B
– Continued –

Find the Greatest Common Factor (GCF) for each of the following, using the method indicated.

6. By listing the factors of each GCF ____________

18 ________________________

45 ________________________

7. By factoring each number as a product of primes GCF ____________

72 ________________________

40 ________________________

Quiz Form A Name ______________________

Class ________ Date ________ Score ______

Unit I – The Structure of Mathematics
Part C – Further Investigation of Operation Symbols

Lesson 1 – Order of Operations
Lesson 2 – Properties of Operations
Lesson 3 – Properties of Operations with Special Numbers

Evaluate the following closed phrases.

1. $10+14\div 7+3$ ______________________

2. $30-3(7)+4$ ______________________

3. $15\div 3+2(7)$ ______________________

4. $48\div 6-36\div 9$ ______________________

5. $36\div 12\cdot 3+4\cdot 2$ ______________________

Name ______________________________

Unit I, Part C, Lessons 1,2, and 3, Quiz Form A
– Continued –

Identify completely the property of operations illustrated by each of the following.

6. $9 \cdot 7 = 7 \cdot 9$ ______________________________

7. $8 + (7 + 5) = (8 + 7) + 5$ ______________________________

8. $x \cdot (a + b) = (x \cdot a) + (x \cdot b)$ ______________________________

9. $4 \cdot (3 \cdot 2) = (4 \cdot 3) \cdot 2$ ______________________________

10. $(x + y) \cdot z = (x \cdot z) + (y \cdot z)$ ______________________________

Identify completely which special property concerning zero or one is used to justify each step in the following problem.

$$[(5 - 5) + 4 \cdot 1] \cdot \frac{1}{4} = n$$

11. $[\ 0 \quad + 4 \cdot 1] \cdot \frac{1}{4} = n$ ______________________________

12. $[\ 0 \quad + \ 4\ ] \cdot \frac{1}{4} = n$ ______________________________

13. $[\quad 4 \quad] \cdot \frac{1}{4} = n$ ______________________________

14. $1 = n$ ______________________________

Quiz Form B

Name ______________________

Class ________ **Date** ________ **Score** ______

Unit I – The Structure of Mathematics
Part C – Further Investigation of Operation Symbols

Lesson 1 – Order of Operations
Lesson 2 – Properties of Operations
Lesson 3 – Properties of Operations with Special Numbers

Evaluate the following closed phrases.

1. $8+14\div 2+3$ ______________________

2. $3(7)+24\div 8$ ______________________

3. $12\div 2+3(4)$ ______________________

4. $24\div 6-10\div 5$ ______________________

5. $40-5\cdot 7+10\div 5$ ______________________

Name ____________________

Unit I, Part C, Lessons 1,2, and 3, Quiz Form B
– Continued –

Identify completely the property of operations illustrated by each of the following.

6. $9+(x+y)=(9+x)+y$ ____________________

7. $8\cdot(3+9)=(8\cdot 3)+(8\cdot 9)$ ____________________

8. $4\cdot 8=8\cdot 4$ ____________________

9. $6\cdot(5\cdot 4)=(6\cdot 5)\cdot 4$ ____________________

10. $7\cdot(5-4)=(7\cdot 5)-(7\cdot 4)$ ____________________

Identify completely which special property concerning zero or one is used to justify each step in the following problem.

$\left(5\cdot\frac{1}{5}\right)+[7\cdot 1+(3-3)]=n$

11. $1+[7\cdot 1+(3-3)]=n$ ____________________

12. $1+[7+(3-3)]=n$ ____________________

13. $1+[7+0]=n$ ____________________

14. $1+[7]=n$ ____________________

$8=n$ **Arithmetic Fact**

Quiz Form A

Name ______________________________

Class ________ **Date** _________ **Score** _____

Unit I – The Structure of Mathematics

Part C – Further Investigation of Operation Symbols

Lesson 4 – Operations with Fractions – Multiplication
Lesson 5 – Operations with Fractions – Addition and Subtraction
Lesson 6 – Operations with Fractions – Division

Use the given rectangle to illustrate the multiplication of the two fractions.

1. $\frac{3}{4} \cdot \frac{1}{5}$

Find the answer to each of the following problems, showing the process of the operation and simplifying your result (if possible). Circle your answer.

2. $\frac{2}{3} \cdot \frac{5}{7} =$

3. $\frac{5}{8} \cdot \frac{6}{7} =$

4. $\frac{12}{35} \cdot \frac{14}{15} =$

5. $\frac{15}{9} \cdot \frac{3}{20} =$

Find the answer to each of the following problems, showing the process of the operation and simplifying your result (if possible). Circle your answer.

6. $\frac{3}{4} + \frac{2}{3} =$

7. $\frac{7}{11} - \frac{1}{3} =$

Name ______________________________

Unit I, Part C, Lessons 4, 5 and 6, Quiz Form A
– Continued –

8. $\frac{2}{9} + \frac{5}{9} =$

9. $\frac{15}{8} - \frac{5}{6} =$

Evaluate each of the following closed phrases by remembering that division means fractions. Be sure to simplify your answer, then circle it.

10. $(5 \div 8) \div (3 \div 4)$

11. $(9 \div 10) \div (2 \div 5)$

Evaluate each of the following by performing the indicated division, remembering the pattern developed in the lesson. Circle your answer.

12. $\frac{5}{6} \div \frac{9}{10}$

13. $\frac{2}{5} \div \frac{4}{7}$

14. $\frac{3}{8} \div \frac{5}{11}$

15. $\frac{4}{9} \div \frac{16}{27}$

Looking back...

16. Simplify: $16 \cdot 30 - (9 + 4)$

17. $(8+7)+4 = 8+(7+4)$ is an illustration of which property of operations? (Identify completely.)

Quiz Form B

Name ______________________

Class ________ Date __________ Score ______

Unit I – The Structure of Mathematics
Part C – Further Investigation of Operation Symbols
Lesson 4 – Operations with Fractions – Multiplication
Lesson 5 – Operations with Fractions – Addition and Subtraction
Lesson 6 – Operations with Fractions – Division

Use the given rectangle to illustrate the multiplication of the two fractions.

1. $\frac{3}{5} \cdot \frac{1}{4}$

Find the answer to each of the following problems, showing the process of the operation and simplifying your result (if possible). Circle your answer.

2. $\frac{2}{5} \cdot \frac{7}{11} =$

3. $\frac{5}{9} \cdot \frac{3}{7} =$

4. $\frac{14}{45} \cdot \frac{27}{56} =$

5. $\frac{17}{11} \cdot \frac{22}{51} =$

Find the answer to each of the following problems, showing the process of the operation and simplifying your result (if possible). Circle your answer.

6. $\frac{5}{8} + \frac{4}{5} =$

7. $\frac{3}{7} - \frac{1}{5} =$

Name ____________________

Unit I, Part C, Lessons 4, 5 and 6, Quiz Form B
– Continued –

8. $\dfrac{3}{11} + \dfrac{7}{11} =$

9. $\dfrac{17}{9} - \dfrac{5}{6} =$

Evaluate each of the following closed phrases by remembering that division means fractions. Be sure to simplify your answer, then circle it.

10. $(7 \div 9) \div (2 \div 3)$

11. $(11 \div 12) \div (3 \div 4)$

Evaluate each of the following by performing the indicated division, remembering the pattern developed in the lesson. Circle your answer.

12. $\dfrac{3}{7} \div \dfrac{14}{15}$

13. $\dfrac{5}{7} \div \dfrac{10}{9}$

14. $\dfrac{5}{8} \div \dfrac{7}{11}$

15. $\dfrac{5}{13} \div \dfrac{20}{39}$

Looking back...

16. Simplify: $(8-2) \cdot 4 + 7$

17. $6 \cdot (4 - c) = (6 \cdot 4) - (6 \cdot c)$ is an illustration of which property of operations? (Identify completely.)

Quiz Form A

Name ______________________

Class ________ Date __________ Score ______

Unit I – The Structure of Mathematics
Part C – Further Investigations of Operation Symbols
Lesson 7 – Operations with Decimals

Evaluate each of the following by performing the indicated operation.

1. $5.371 + 7.35$ ____________________

2. $15.79 - 8.53$ ____________________

3. $17.2 - 9.86$ ____________________

4. $32.98 \cdot 4.3$ ____________________

Name ______________________________

Unit I, Part C, Lesson 7, Quiz Form A
– Continued –

5. $2.1 \div .05$ ____________________

6. $27.48 \div 4$ ____________________

7. $13.3 \cdot .005$ ____________________

8. $10.608 \div 4.42$ ____________________

Quiz Form B

Name ______________________

Class __________ Date __________ Score ______

Unit I – The Structure of Mathematics
Part C – Further Investigation of Operation Symbols
Lesson 7 - Operations with Decimals

Evaluate each of the following by performing the indicated operation.

1. $.8 \cdot 7.3$ ____________________

2. $.8149 + 6.74$ ____________________

3. $.94882 \div 6.6$ ____________________

4. $14.21 - 7.8941$ ____________________

Name ______________________________

Unit I, Part C, Lesson 7, Quiz Form B
– Continued –

5. $.86 \div 43$ ____________

6. $13.648 + .293$ ____________

7. $7.54 \cdot 1.328$ ____________

8. $15 - 2.64$ ____________

Quiz Form A Name ______________________

Class __________ Date __________ Score ________

Unit I – The Structure of Mathematics
Part C – Further Investigation of Operation Symbols
Lesson 8 – Operations with Signed Numbers – Vectors and Absolute Value
Lesson 9 – Operations with Signed Numbers – Addition
Lesson 10 – Operations with Signed Numbers – Subtraction

Tell what signed number is represented by each of the following vectors.

1. 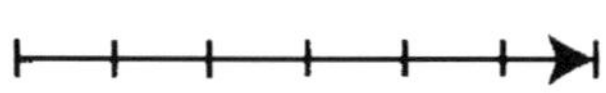

2. 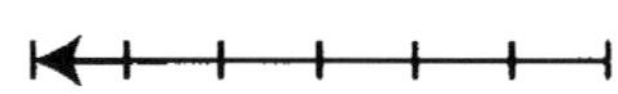

Find the absolute value for each of the following as indicated.

3. $\left| {}^{+}17 \right|$

4. $\left| {}^{-}7 \right|$

5. $\left| 5.3 \right|$

Find the answer for the following addition problem, using the given number line and vector addition.

6. ${}^{+}5 + {}^{+}2 + {}^{-}4$

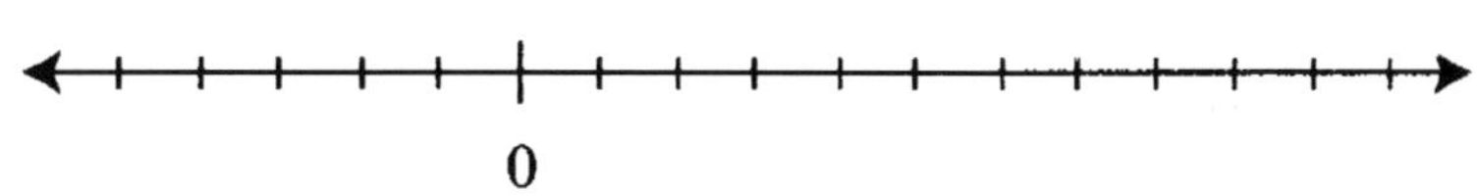

Find the answer mentally for each of the following problems.

7. ${}^{+}8 + {}^{+}3$

8. ${}^{+}9 + {}^{-}4$

9. ${}^{-}3 + {}^{-}6$

Find the answer for each of the following problems by first <u>rewriting</u> the problem as a sum of signed numbers.

10. $9 - 7$

11. $24 - 36$

12. ${}^{-}7 - {}^{+}8$

Quiz Form B

Name ____________________

Class ________ Date ________ Score ______

Unit I – The Structure of Mathematics

Part C – Further Investigation of Operation Symbols

Lesson 8 – Operations with Signed Numbers – Vectors and Absolute Value

Lesson 9 – Operations with Signed Numbers – Addition

Lesson 10 – Operations with Signed Numbers – Subtraction

Tell what signed number is represented by each of the following vectors.

1. 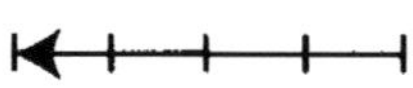

2. 

Find the absolute value for each of the following as indicated.

3. $|^{-}13|$

4. $|6.4|$

5. $|^{+}1|$

Find the answer for the following addition problem, using the given number line and vector addition.

6. $^{+}6 + ^{-}4 + ^{+}3$

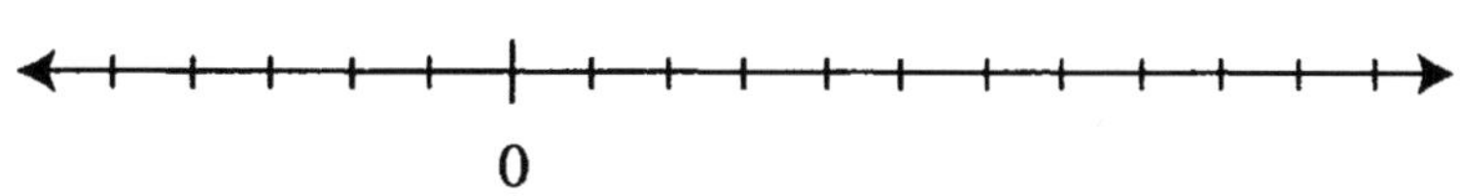

Find the answer mentally for each of the following problems.

7. $^{-}5 + ^{-}7$

8. $^{+}9 + ^{+}8$

9. $^{+}11 + ^{-}4$

Find the answer for each of the following problems by first rewriting the problem as a sum of signed numbers.

10. $^{-}6 - ^{-}7$

11. $12 - 9$

12. $^{-}6 - ^{+}10$

Quiz Form A

Name ______________________

Class ________ Date ________ Score ______

Unit I – The Structure of Mathematics
Part C – Further Investigation of Operation Symbols
Lesson 11 – Operations with Signed Numbers – Multiplication and Division

Perform the indicated operation in each of the following. Be careful to attach the proper sign to your answer according to the guidelines of this lesson.

1. $^{+}6 \cdot {}^{+}21$ ______________

2. $^{-}13 \cdot {}^{-}32$ ______________

3. $^{+}64 \div {}^{-}32$ ______________

4. $^{-}8 \cdot {}^{-}7$ ______________

5. $^{-}32 \div {}^{+}4$ ______________

6. $^{-}31 \cdot {}^{+}11$ ______________

7. $^{+}7 \cdot {}^{-}15$ ______________

8. $^{+}81 \div {}^{+}27$ ______________

9. $^{-}72 \div {}^{-}9$ ______________

10. $^{+}4 \div {}^{-}4$ ______________

Quiz Form B

Name ______________________

Class __________ Date __________ Score ______

Unit I – The Structure of Mathematics
Part C – Further Investigation of Operation Symbols
Lesson 11 – Operations with Signed Numbers – Multiplication and Division

---

Perform the indicated operation in each of the following. Be careful to attach the proper sign to your answer according to the guidelines of this lesson.

1. $^{-}8 \div {}^{+}8$ ____________

2. $^{-}64 \div {}^{-}16$ ____________

3. $^{+}36 \div {}^{+}12$ ____________

4. $^{+}8 \cdot {}^{-}17$ ____________

5. $^{-}21 \cdot {}^{+}13$ ____________

6. $^{-}44 \div {}^{+}11$ ____________

7. $^{-}6 \cdot {}^{-}8$ ____________

8. $^{+}81 \div {}^{-}27$ ____________

9. $^{-}14 \cdot {}^{-}12$ ____________

10. $^{+}7 \cdot {}^{+}31$ ____________

Quiz Form A | **Name** ______________________________

Class ________ **Date** __________ **Score** ______

Unit I – The Structure of Mathematics
Part D – Further Investigation of Relation Symbols
Lesson 1 – Order of Numbers and the Number Line

For each of the following, label the appropriate points on the given number line by assigning the given numbers to those points as indicated.

1. $^{+}5$ assigned to point C

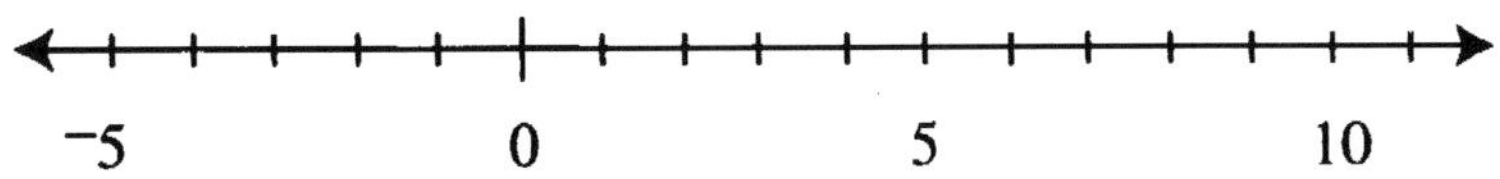

2. $^{-}4$ assigned to point T

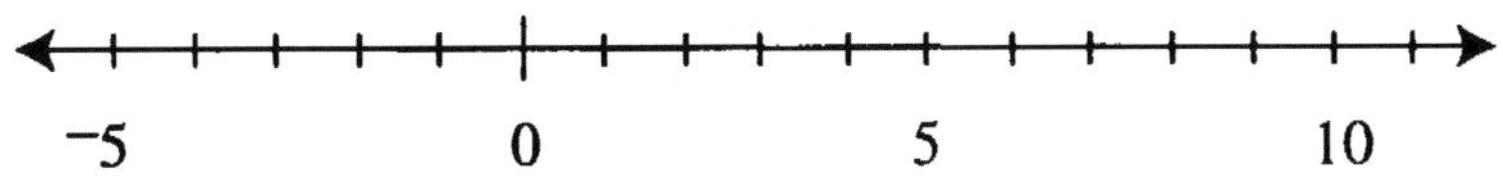

3. $^{-}3\frac{3}{4}$ assigned to point U

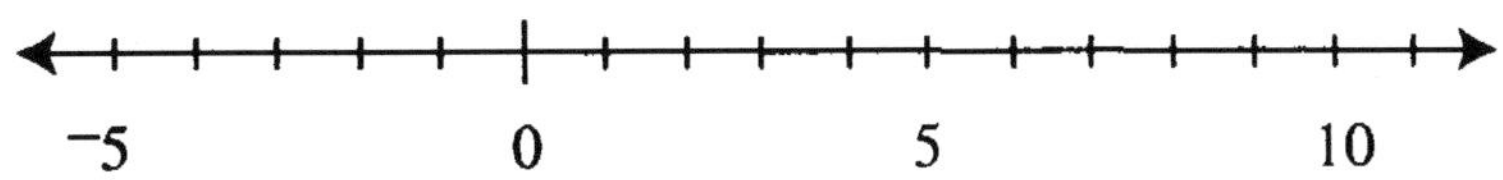

4. $^{+}1.75$ assigned to point X

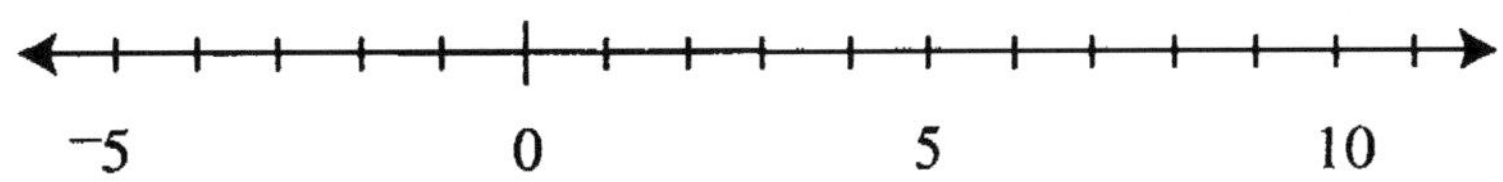

5. $^{-}2.2$ assigned to point A

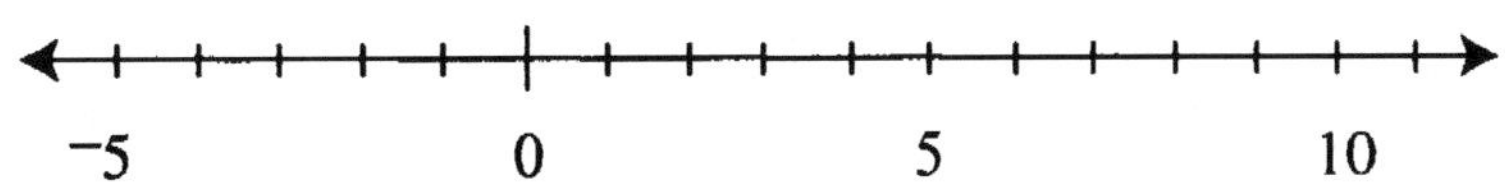

Use one of the three basic relation symbols (=, >, <) in each of the following to make a true statement of relationship.

6. $^{-}5$ _____ 0

7. $^{-}9$ _____ $^{-}5$

8. 0 _____ $^{+}3$

9. $^{+}4$ _____ $^{-}4$

10. $\frac{^{-}3}{7}$ _____ $\frac{2}{9}$

11. $^{+}9$ _____ $^{+}5$

12. $.05$ _____ $.5$

13. $\frac{1}{3}$ _____ $\frac{1}{4}$

14. $^{-}7.2$ _____ $^{-}1.5$

15. If $^{+}5$ dollars means 5 dollars deposited in (or put into) a bank account, what does $^{-}5$ dollars mean?

Quiz Form B Name ______________________

Class ________ Date __________ Score ______

Unit I – The Structure of Mathematics
Part D – Further Investigation of Relation Symbols
Lesson 1 – Order of Numbers and the Number Line

For each of the following, label the appropriate points on the given number line by assigning the given numbers to those points as indicated.

1. $^{-}7$ assigned to point C

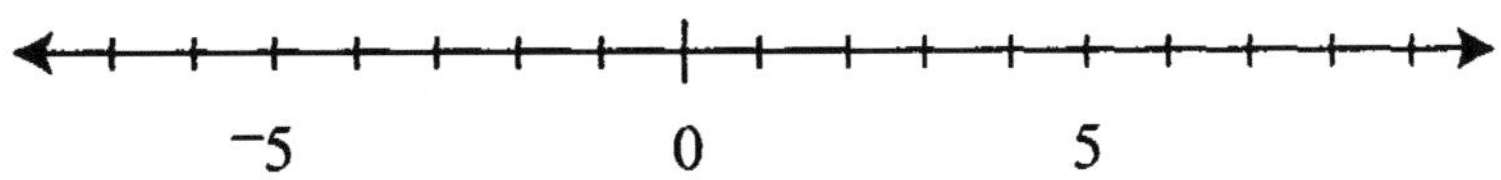

2. $^{+}3$ assigned to point T

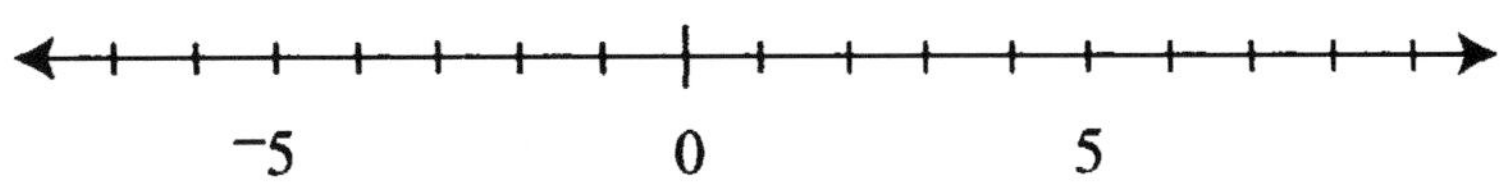

3. $4\frac{1}{4}$ assigned to point U

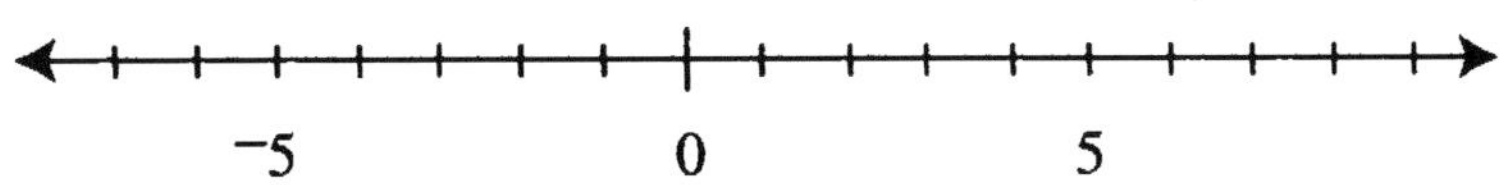

4. $^{-}1.25$ assigned to point X

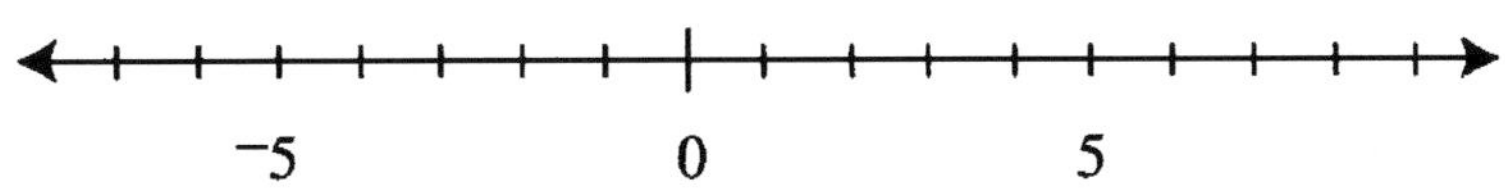

5. $^{-}3.4$ assigned to point A

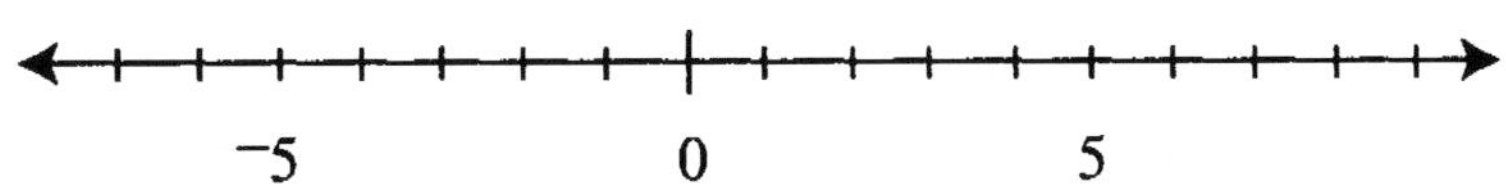

Use one of the three basic relation symbols (=, >, <) in each of the following to make a true statement of relationship.

6. $^{+}4$ _____ 0

7. $^{-}7$ _____ $^{-}11$

8. $^{+}2$ _____ $^{+}6$

9. $\frac{2}{11}$ _____ $\frac{^{-}3}{7}$

10. $\frac{1}{6}$ _____ $\frac{1}{5}$

11. .7 _____ .07

12. 0 _____ $^{-}7$

13. $^{-}6.1$ _____ $^{-}1.3$

14. 8 _____ $^{-}8$

15. If $^{-}2$ means 2 miles below sea level, what does $^{+}2$ miles mean?

Quiz Form A Name ____________

Class ________ Date ________ Score ______

Unit I – The Structure of Mathematics
Part D – Further Investigation of Relation Symbols
Lesson 2 – Properties of Equality
Lesson 3 – Properties of Inequality

State which property of equality or inequality justifies each of the following statements.

1. If $12 = 3x$, then $3x = 12$ ____________

2. If $y + 4 = 9$, and $9 = 5 + 4$, then $y + 4 = 5 + 4$ ____________

3. $8 - 7xy = 8 - 7xy$ ____________

4. If $2 > {}^{-}1$, then $2 + 3 > {}^{-}1 + 3$ ____________

5. If $x = y$, then $x - a = y - a$ ____________

6. If $3a = 9b$, then $a = 3b$ ____________

7. ${}^{-}6x > 12$ is the same as $x < {}^{-}2$ ____________

In each of the following, perform the indicated operation on the given relation, indicating what the resulting sentence is.

8. $24 - 5 = 17 + 2$

Add ${}^{-}2$

9. $10 > 2$

Mult $\frac{{}^{-}1}{2}$

Name ____________________

Unit I, Part D, Lessons 2 and 3, Quiz Form A
– Continued –

Extending your mind...

A relation R on a set A is an equivalence relation if it has the following properties.

Reflexive Property	For any element *a* of set A, *a* R *a*.
Symmetric Property	For all elements *a* and *b* of set A, if *a* R *b*, then *b* R *a*.
Transitive Property	For all elements *a*, *b*, and c of set A, if *a* R *b* and *b* R *c*, then *a* R *c*.

Equality on the set of all real numbers is reflexive, symmetric, and transitive. Therefore, it is an equivalence relation.

In each of the following, a relation and a set are given. Write yes if the relation is an equivalence relation on the given set. If it is not, write no, and tell which of the properties fail to hold true.

10. > {all numbers} ____________________

11. "is the brother of" {all boys in your school} ____________________

Quiz Form B

Name ____________________

Class ________ Date ________ Score ______

Unit I – The Structure of Mathematics
Part D – Further Investigation of Relation Symbols
Lesson 2 – Properties of Equality
Lesson 3 – Properties of Inequality

State which property of equality or inequality justifies each of the following statements.

1. If $5 < 7$, then $5 + 4 < 7 + 4$ ____________________

2. $17 + 3x = 17 + 3x$ ____________________

3. If $^{-}3x < 12$, then $x > {}^{-}4$ ____________________

4. If $7 = 8 + {}^{-}1$, and $8 + {}^{-}1 = 21 \div 3$, then $7 = 21 \div 3$ ____________________

5. If $21 = 3x$, then $3x = 21$ ____________________

6. If $4x = 20$, then $x = 5$ ____________________

7. If $a = b$, then $a + x = b + x$ ____________________

In each of the following, perform the indicated operations on the given relation, indicating what the resulting sentence is.

8. $14 > 7$

Mult $\frac{^{-}1}{7}$

9. $19 + 7 = 25 + 1$

Subtract 1

Name ______________________________

Unit I, Part D, Lessons 2 and 3, Quiz Form B
- Continued -

Extending your mind...

A relation R on a set A is an equivalence relation if it has the following properties.

Reflexive Property	For any element a of set A, a R a.
Symmetric Property	For all elements a and b of set A, if a R b, then b R a.
Transitive Property	For all elements a, b, and c of set A, if a R b and b R c, then a R c.

Equality on the set of all real numbers is reflexive, symmetric, and transitive. Therefore, it is an equivalence relation.

In each of the following, a relation and a set are given. Write yes if the relation is an equivalence relation on the given set. If it is not, write no, and tell which of the properties fail to hold true.

10. "$\neq$" {all numbers} ____________

11. "graduates the same year as" {any student in your school} ____________

Quiz Form A Name ____________________

Class ________ Date __________ Score ______

Unit I – The Structure of Mathematics
Part E – Mathematical Models
Lesson 1 – The Mathematics of Sets

Given the following sets, find the union or intersection as indicated.

$A = \{1, 2, 3, 4, 5, 6, 7, ... 16, 17, 18\}$ $B = \{2, 4, 6, 8\}$

$C = \{3, 4, 5, 6\}$ $D = \{1, 2, 17, 18\}$

1. $C \cup D$ ____________________

2. $B \cup C$ ____________________

3. $D \cap B$ ____________________

4. $C \cap B$ ____________________

Using the given sets for exercises 1-4, indicate if the following are true or false.

5. $C \subset A$ ____________________

6. $(D \cap B) \subset C$ ____________________

7. $(B \cup C \cup D) \subset A$ ____________________

8. $D \cap C$ is the null set ____________________

9. All of the given sets (A, B, C, and D) could be classified as finite sets ____________________

Name ______________________________

Unit I, Part E, Lesson 1, Quiz Form A
– Continued –

10. If $A = \{a, 2\}$, $B = \{2, 4, c, d\}$ and $A \cap B = \{2\}$, draw a Venn Diagram that would represent the relationship between these sets.

Quiz Form B

Name ______________________

Class ________ Date __________ Score ______

Unit I – The Structure of Mathematics
Part E – Mathematical Models
Lesson 1 - The Mathematics of Sets

Given the following sets, find the union or intersection as indicated.

$A = \{a, b, c, d, e, f, g, h, i, j, k, l, m, n, o, p, q\}$ $B = \{b, d, f, h\}$

$C = \{c, d, e, f\}$ $D = \{a, b, q\}$

1. $C \cup D$ ______________________

2. $B \cup C$ ______________________

3. $D \cap B$ ______________________

4. $C \cap B$ ______________________

Using the given sets for exercises 1-4, indicate if the following are true or false.

5. $C \subset A$ ______________________

6. $(D \cap B) \subset C$ ______________________

7. A is finite ______________________

8. $(B \cup C) \cap D$ is the null set ______________________

9. $(C \cup D) \cap B$ contains one element ______________________

Name ________________________________

Unit I, Part E, Lesson 1, Quiz Form B
– Continued –

10. $A = \{x, 5\}$, $B = \{5, 4, m, n\}$ and $A \cap B = \{5\}$, draw a Venn Diagram that would represent the relationship between these sets.

Quiz Form A Name ______________________

Class ________ Date __________ Score ______

Unit I – The Structure of Mathematics
Part E – Mathematical Models
Lesson 2 – The Mathematics of Functions

1. Given a domain $D = \{1, 3, 5, 7\}$ and a range $R = \{3, 7, 11, 15\}$, write an open phrase which defines a function.

For each of the following functions, use the given domain, and find the range.

2. $f(\ ) = 3(\ )$ the domain is $D = \{^{-}1, {}^{-}3, {}^{-}5, {}^{-}7\}$

   so the range is $R = \{\qquad\qquad\}$

3. $f(\ ) = \dfrac{(\ ) - 2}{3}$ the domain is $D = \{5, 8, 11, 14\}$

   so the range is $R = \{\qquad\qquad\}$

4. True or false?

   Using the definitions in this lesson, it could be said that all functions are relations, but all relations are not functions. Explain your answer.

Quiz Form B

Name ______________________

Class ________ Date ________ Score ______

Unit I – The Structure of Mathematics
Part E – Mathematical Models
Lesson 2 – The Mathematics of Functions

1. Given a domain $D = \{2, 4, 6, 8\}$ and a range $R = \{7, 11, 15, 19\}$, write an open phrase which defines a function.

For each of the following functions, use the given domain, and find the range.

2. $f(\ \ ) = {}^{-}2(\ \ )$ the domain is $D = \{{}^{-}1, {}^{-}3, {}^{-}5, {}^{-}7\}$

so the range is $R = \{ \qquad \}$

3. $f(\ \ ) = \frac{(\ \ ) + 2}{4}$ the domain is $D = \{10, 14, 18, 22\}$

so the range is $R = \{ \qquad \}$

4. True or false?

Using the definitions in this lesson, if the domain was D = {all women in the world who are mothers}, and the range was R = {all women in the world}, then the relation defined by pairing each mother in the world with their daughter(s) would define a function. Explain your answer.

Unit Test Form A Name ______________________

Class ________ Date __________ Score ______

Unit I – The Structure of Mathematics

In problems 1-5, state the five parts of mathematical speech, and give an example of each.

1. Example:

2. Example:

3. Example:

4. Example:

5. Example:

For each expression in problems 6-10, tell (a) whether the expressions are open or closed, sentences or phrases, and (b) do whatever you can with each. If you need a domain, use $\{1, 2, 3\}$. If you need a replacement set, use $\{4, 5, 6\}$.

6. $m+11$ (a). (b).

7. $9\cdot 7+0$ (a). (b).

8. $3=6-x$ (a). (b).

9. 27,613 (a). (b).

10. $\dfrac{30-12}{3}=7-2$ (a). (b).

11. Tell which of the following are rational numbers by circling the appropriate numbers.

$\sqrt{29}$ π $^{-}15$ $\dfrac{6}{7}$ $5.\overline{2}$ $\sqrt{36}$

Name ______________________________

Unit I – The Structure of Mathematics, Unit Test Form A
– Continued, Page 2 –

12. Express $\frac{3}{16}$ as a decimal. Show your work.

13. Complete the following to find the equivalent form.

$$\frac{3}{8} = \frac{}{72}$$

14. Simplify the following by writing both the numerator and the denominator in prime factored form, and then finding all the "1's" you can.

$$\frac{42}{90}$$

15. Express $\frac{1}{11}$ as a decimal. Show your work.

16. Express .45 as a fraction. Show your work.

17. Express the number 1.2 as a percent.

18. Express $.\overline{39}$ as a fraction. Show your work.

For problems 19-21, factor each number as a product of primes. Then, in problem 22, find the greatest common factor of the three numbers.

19. 18

20. 30

21. 42

22. The greatest common factor is:

Name ______________________________

Unit I – The Structure of Mathematics, Unit Test Form A
– Continued, Page 3 –

Simplify each of the expressions in problems 23 and 24.

23. $(^{-}12)(5)-6$

24. $62\cdot 8-(5-64\div 4)$

For problems 25-27, find $-a$ when a is given. (In other words, find the opposite, or additive inverse, of each number.)

25. $a={}^{-}21$

26. $a=8$

27. $a=\frac{5}{b}$

Rewrite each expression in problems 28 and 29 using the indicated property of operations.

28. $m\cdot p$ Commutative Law

29. $7+(a+b)$ Associative Law

Find each sum in problems 30-32.

30. $17+{}^{-}24$

31. ${}^{-}3.6+{}^{-}7.8$

32. $\frac{^{-}6}{7}+\frac{1}{2}$

Name ______________________________

Unit I – The Structure of Mathematics, Unit Test Form A – Continued, Page 4 –

Find each product in problems 33-35.

33. $^-3(2)(^-1)(^-6)$

34. $^-15(^-4.5)$

35. $\frac{7}{8}\left(\frac{^-3}{5}\right)$

Find each difference in problems 36-38.

36. $^-17.1 - 6.3$

37. $^-9 - (^-3)$

38. $\frac{7}{4} - \left(\frac{^-3}{2}\right)$

Find each quotient in problems 39-41.

39. $35.7 \div (^-8.5)$

40. $\frac{11}{12} \div \left(\frac{^-4}{5}\right)$

41. $\frac{^-6.6}{11}$

Name ______________________________

Unit I – The Structure of Mathematics, Unit Test Form A – Continued, Page 5 –

Simplify each of the expressions in problems 42-44.

42. $|48|$

43. $|^{-}4.18|$

44. $|18-18|$

Evaluate and simplify each expression in problems 45 and 46.

45. $6a-2b$ for $a=3$ and $b=1$.

46. $\frac{x-y}{3}$ for $x=17$ and $y=5$.

For problems 47 and 48, write a true statement, using " < " or " > " to replace ☐.

47. $^{-}8$ ☐ 6

48. $^{-}5$ ☐ $^{-}2$

Use the accompanying number lines in problems 49 and 50 to show each of the following.

49. $x={}^{-}2$

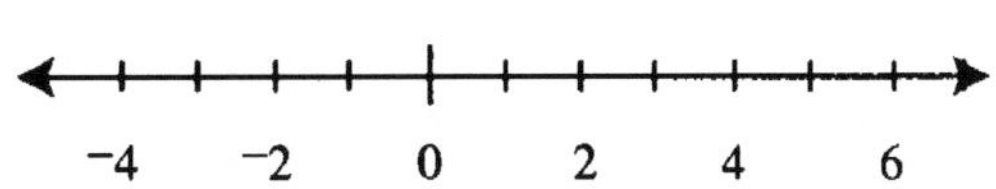

50. Add $5+{}^{-}2$ using vectors.

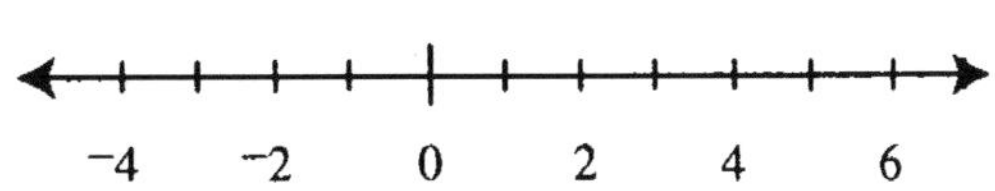

Name ______________________________

Unit I – The Structure of Mathematics, Unit Test Form A
– Continued, Page 6 –

For each of the true statements in problems 51-54, state which property of equality or inequality is illustrated.

51. If $9 = 6 + 3$, then $6 + 3 = 9$

52. $^{-}7 = {^{-}7}$

53. If $x > y$, then $x + a > y + a$

54. If $^{-}12 < 4$, then $3 > {^{-}1}$

55. Find the union of $\{1, 2, 3, 4\}$ and $\{3, 4, 5, 6\}$.

56. Find the intersection of $\{1, 2, 3, 4\}$ and $\{3, 4, 5, 6\}$.

57. Given a function defined as $f(\quad) = {^{-}7}(\quad) + 1$, and a domain $\text{D} = \{3, 5, 7, 9\}$, write an explanation of the process you would use to find the range. Then, actually find the range.

Unit Test Form B

Name ______________________

Class ________ Date ________ Score ______

Unit I – The Structure of Mathematics

In problems 1-4, state the four types of mathematical expressions, and give an example of each.

1. Example:

2. Example:

3. Example:

4. Example:

In problems 5-10, translate each of the English phrases into mathematical expressions.

5. 85% of x

6. three times d

7. 19 more than c

8. 11 less than m

9. twice y

10. the square of a, divided by 8

11. Tell which of the following are irrational numbers by circling the appropriate numbers.

$\sqrt{121}$ $\qquad \pi$ $\qquad \frac{^{-}4}{3}$ $\qquad 9.\overline{7}$ $\qquad ^{-}15$ $\qquad \sqrt{6}$

Name ______________________________

Unit I – The Structure of Mathematics, Unit Test Form B
– Continued, Page 2 –

12. Complete the following to find the equivalent form.

$$\frac{11}{3}=\frac{\quad}{45}$$

13. Simplify the following by writing both the numerator and the denominator in prime factored form, and then finding all the "1's" you can.

$$\frac{30}{105}$$

14. Express $\frac{3}{8}$ as a decimal. Show your work.

15. Express $\frac{3}{11}$ as a decimal. Show your work.

16. Express .52 as a fraction. Show your work.

17. Express $.\overline{21}$ as a fraction. Show your work.

18. Express the number .14 as a percent.

For problems 19-21, factor each number as a product of primes. Then, in problem 22, find the least common multiple of the three numbers.

19. 9

20. 6

21. 15

22. The least common multiple is:

Name ______________________________

Unit I – The Structure of Mathematics, Unit Test Form B – Continued, Page 3 –

Simplify each of the expressions in problems 23 and 24.

23. $53 - 4 \cdot 28 + 34$

24. $(6)\left({}^{-}13\right) \div 3$

For problems 25-27, find $-a$ when a is given. (In other words, find the opposite, or additive inverse, of each number.)

25. $a = 5$

26. $a = {}^{-}24$

27. $a = 6.8$

Rewrite each expression in problems 28 and 29 using the indicated property of operations.

28. $d + (e + f)$ Associative Law

29. $a + b$ Commutative Law

Find each sum in problems 30-32.

30. $\frac{{}^{-}5}{7} + \frac{1}{3}$

31. ${}^{-}4.7 + \left({}^{-}5.2\right)$

32. $15 + {}^{-}23$

Name ____________________________

Unit I – The Structure of Mathematics, Unit Test Form B – Continued, Page 4 –

Find each product in problems 33-35.

33. $(^-4.1)(^-1.2)$

34. $\frac{^-2}{3}\left(\frac{6}{7}\right)$

35. $^-6(4)(^-3)(5)$

Find each difference in problems 36-38.

36. $^-40.1 - 6.8$

37. $^-17 - (^-5)$

38. $\frac{8}{9} - \left(\frac{^-2}{3}\right)$

Find each quotient in problems 39-41.

39. $\left(\frac{^-7}{3}\right) \div \frac{5}{6}$

40. $8.64 \div (^-3.6)$

41. $\frac{^-55.8}{^-6.2}$

Name ______________________________

Unit I – The Structure of Mathematics, Unit Test Form B
– Continued, Page 5 –

Simplify each of the expressions in problems 42-44.

42. $\left| \frac{^-3}{11} \right|$

43. $|26-26|$

44. $|17.2|$

Evaluate and simplify each expression in problems 45 and 46.

45. $\frac{5a}{2b}$ for $a=2$ and $b=3$.

46. $\frac{4x-y}{3}$ for $x=5$ and $y=2$.

For problems 47 and 48, write a true statement using " < " or " > " to replace ☐.

47. 5 ☐ $^-4$

48. $^-7$ ☐ $^-13$

Use the accompanying number lines in problems 49 and 50 to show each of the following.

49. $x=4$

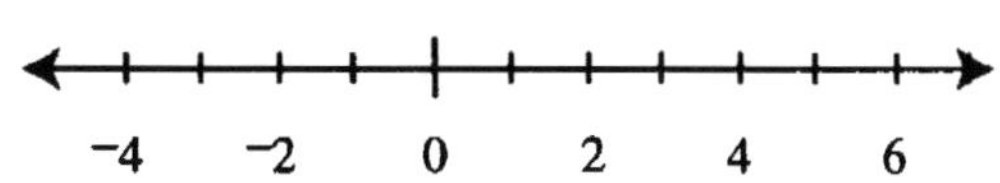

50. Add $^-3+5$ using vectors.

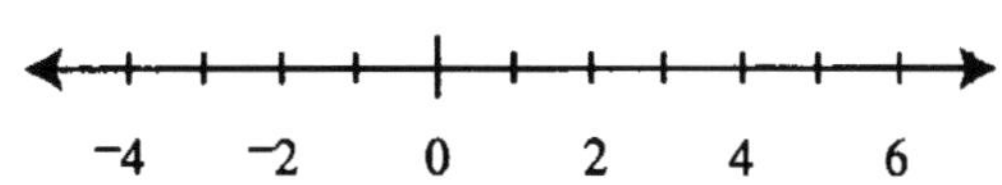

Name ______________________________

Unit I – The Structure of Mathematics, Unit Test Form B
– Continued, Page 6 –

For each of the true statements in problems 51-54, state which property of equality or inequality is illustrated.

51. If $a = b$, then $a + c = b + c$

52. If $a = b$, then $b = a$

53. If $12 > 4$, then $3 > 1$

54. If $5 = 4 + 1$, and $4 + 1 = 10 \div 2$, then $5 = 10 \div 2$

55. Find the intersection of $\{a, b, c, d\}$ and $\{b, c, e, f\}$.

56. Find the union of $\{a, b, c, d\}$ and $\{b, c, e, f\}$.

57. Given a function defined as $f(\ \ ) = 5(\ \ ) - 2$, and a domain $D = \{3, 5, 7, 9\}$, write an explanation of the process you would use to find the range. Then actually find the range.